O9-ABI-889

Jr. Graphic Environmental Dangers™

Global Warming

Greenhouse Gases and the Ozone Layer

Daniel R. Faust

PowerKiDS press.
New York

Published in 2009 by The Rosen Publishing Group, Inc.
29 East 21st Street, New York, NY 10010

First Edition

Editors: Joanne Randolph
Book Design: Greg Tucker
Illustrations: Dheeraj Verma/Edge Entertainment

Library of Congress Cataloging-in-Publication Data

Faust, Daniel R.
Global warming : greenhouse gases and the ozone layer / Daniel R. Faust. — 1st ed.
p. cm. — (Jr. graphic environmental dangers)
Includes index.
ISBN 978-1-4042-4260-9 (library binding) — ISBN 978-1-4042-4599-0 (pbk.) —
ISBN 978-1-4042-3984-5 (6-pack)
1. Global warming—Juvenile literature. 2. Greenhouse effect, Atmospheric—Juvenile literature. 3. Ozone layer depletion—Juvenile literature. I. Title.
QC981.8.G56F38 2009
363.738'74—dc22

2007049817

Manufactured in the United States of America

CONTENTS

INTRODUCTION

Earth is in trouble and it needs our help. **Temperatures** on Earth are rising, and over time this will lead to major problems for people, plants, and animals. One reason temperatures are rising is because greenhouse gases are increasing. Temperatures are also rising because there is a hole in the ozone layer, which keeps most of the Sun's rays from reaching Earth.

What will happen if we cannot get global warming under control or close the hole in the ozone layer? Read on and find out!

GLOBAL WARMING

GREENHOUSE GASES AND THE OZONE LAYER

THE SUN.

ALL LIFE ON EARTH DEPENDS ON THE ENERGY GENERATED FROM THE SUN IN ONE WAY OR ANOTHER. THE SUN PROVIDES OUR PLANET WITH LIGHT AND WARMTH.

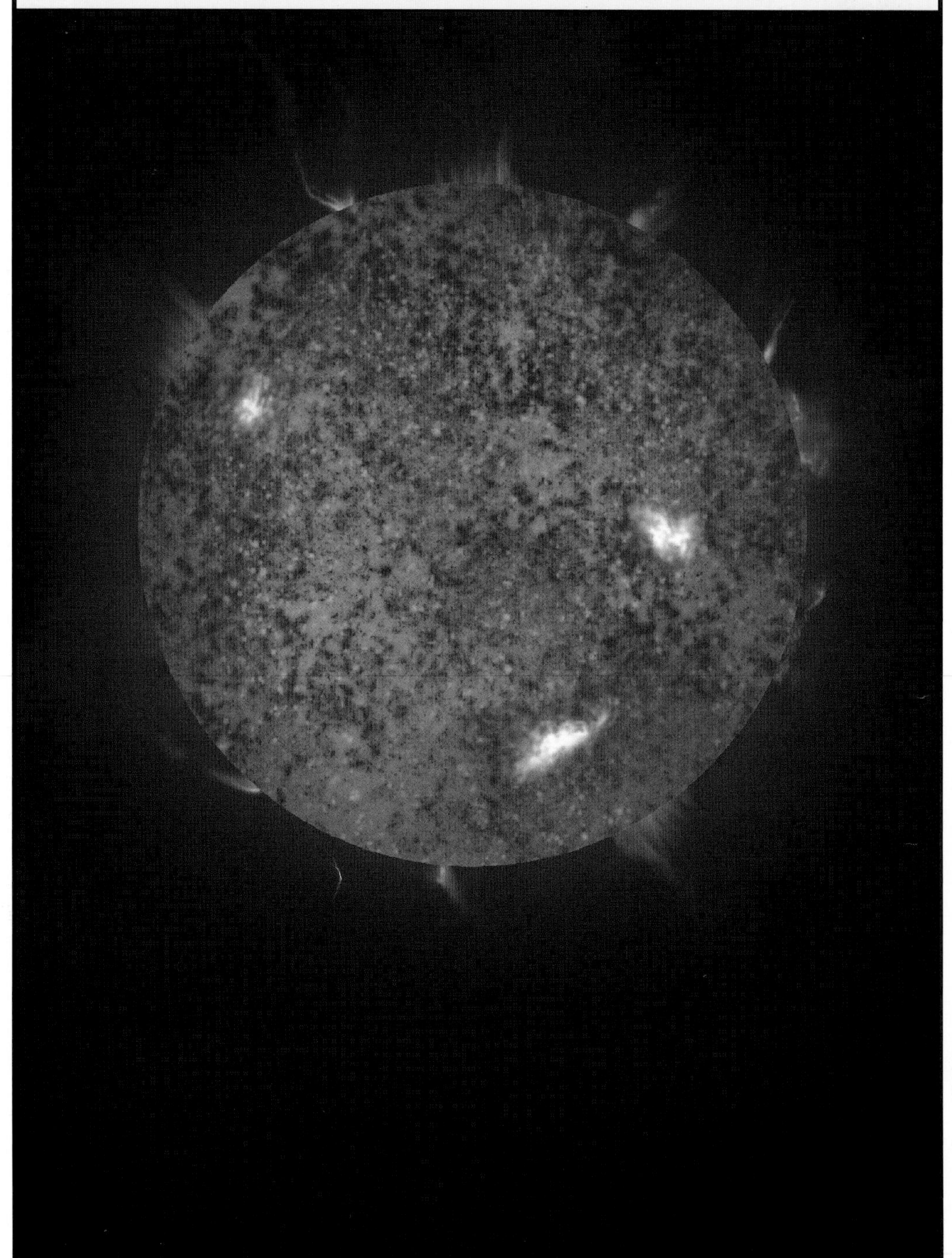

HOWEVER, TOO MUCH **SOLAR RADIATION** CAN CAUSE SERIOUS PROBLEMS.

EARTH IS SURROUNDED BY AN **ATMOSPHERE**. THIS ATMOSPHERE IS LIKE A BLANKET AROUND EARTH THAT IS MADE OF MANY DIFFERENT GASES.

SOME OF THESE GASES, LIKE OXYGEN AND CARBON DIOXIDE, ARE NEEDED FOR PLANTS AND ANIMALS TO LIVE.

EXOSPHERE

THERMOSPHERE

MESOSPHERE

STRATOSPHERE

TROPOSPHERE

LAYERS OF EARTH'S ATMOSPHERE

OZONE IS A GAS THAT IS MADE UP OF THREE **ATOMS** OF OXYGEN.

THE OXYGEN WE BREATHE IS MADE UP OF TWO ATOMS OF OXYGEN. THE ATMOSPHERE HAS BOTH OF THESE GASES, AS WELL AS SINGLE ATOMS OF OXYGEN.

O

OXYGEN ATOMS

OXYGEN MOLECULE

O_2

O_3

OZONE MOLECULE

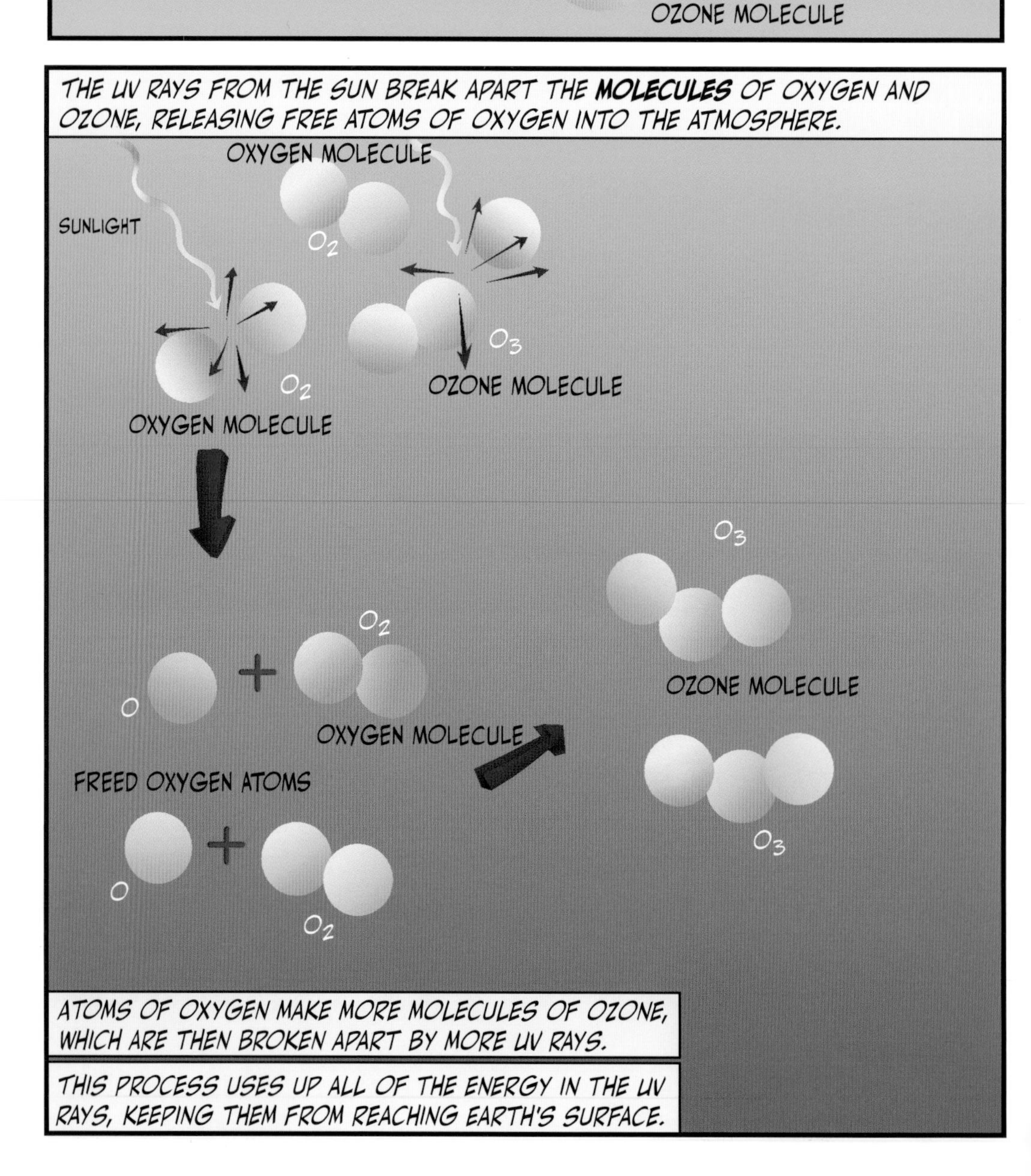

HOWEVER, NOT ALL UV RAYS ARE STOPPED THIS WAY.
SOME UV RAYS STILL MAKE IT TO EARTH'S SURFACE.

HEY GUYS, WE'D BETTER GO PUT ON MORE SUNSCREEN.
I'M NOT READY TO LEAVE THE WATER YET. BESIDES, SUNSCREEN IS FOR KIDS. I'M TRYING TO GET A TAN.

SARA'S RIGHT, WE ALL NEED TO PUT ON MORE SUNSCREEN. WE SHOULD PROBABLY WEAR OUR HATS, TOO.

YOU GUYS ARE SO LAME.

IT'S NOT LAME, TOMMY. IT'S REALLY IMPORTANT. DO YOU REMEMBER THE SCARE MY DAD WENT THROUGH WITH SKIN CANCER?
HE SAYS IT WOULD NOT HAVE HAPPENED IF HE HAD JUST WORN SUNSCREEN ALL THE TIME.
SUNSCREEN HELPS KEEP OUT THE UV RAYS.

WHAT ARE UV RAYS?

WEREN'T YOU PAYING ATTENTION IN CLASS LAST WEEK? THE SUN PRODUCES HEAT AND LIGHT, BUT IT ALSO PRODUCES OTHER KINDS OF ENERGY.
THAT'S RIGHT. AND ONE KIND OF ENERGY IS ULTRAVIOLET RADIATION*, OR UV RAYS.
*RADIATION IS ENERGY THAT IS RADIATED, OR SENT OUT, IN THE FORM OF RAYS, WAVES, OR **PARTICLES**.

"UV RAYS ARE WHAT GIVE PEOPLE SUNTANS. TANS ARE NOT HEALTHY, THOUGH. A TAN IS A SIGN THAT THE SUN HAS HURT YOUR SKIN.
"UV RAYS CAN ALSO CAUSE PAINFUL SUNBURNS.

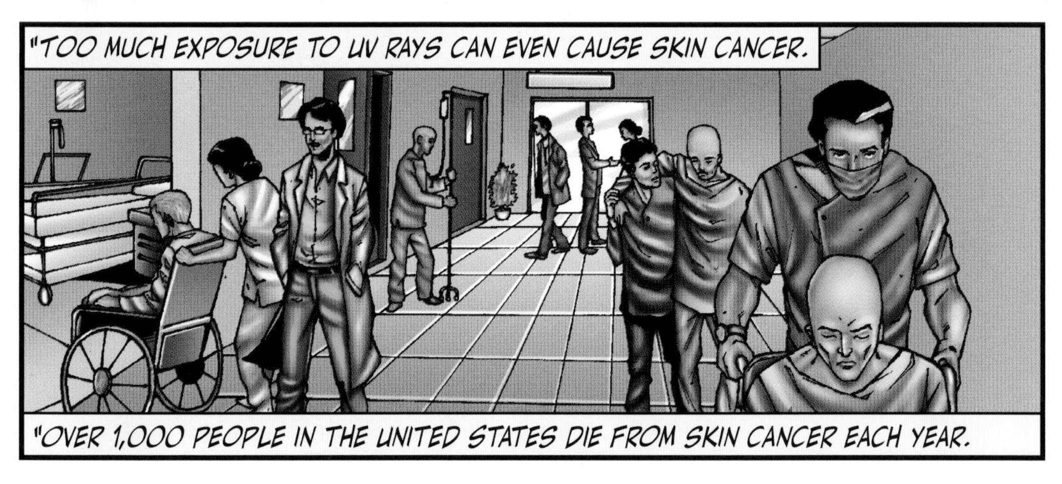
"TOO MUCH EXPOSURE TO UV RAYS CAN EVEN CAUSE SKIN CANCER.
"OVER 1,000 PEOPLE IN THE UNITED STATES DIE FROM SKIN CANCER EACH YEAR.

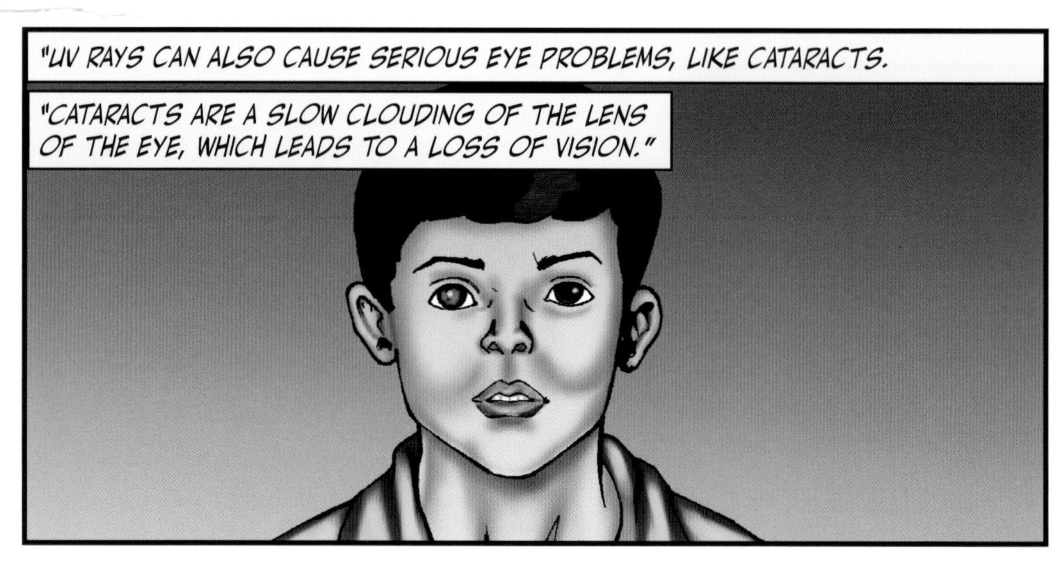
"UV RAYS CAN ALSO CAUSE SERIOUS EYE PROBLEMS, LIKE CATARACTS.
"CATARACTS ARE A SLOW CLOUDING OF THE LENS OF THE EYE, WHICH LEADS TO A LOSS OF VISION."

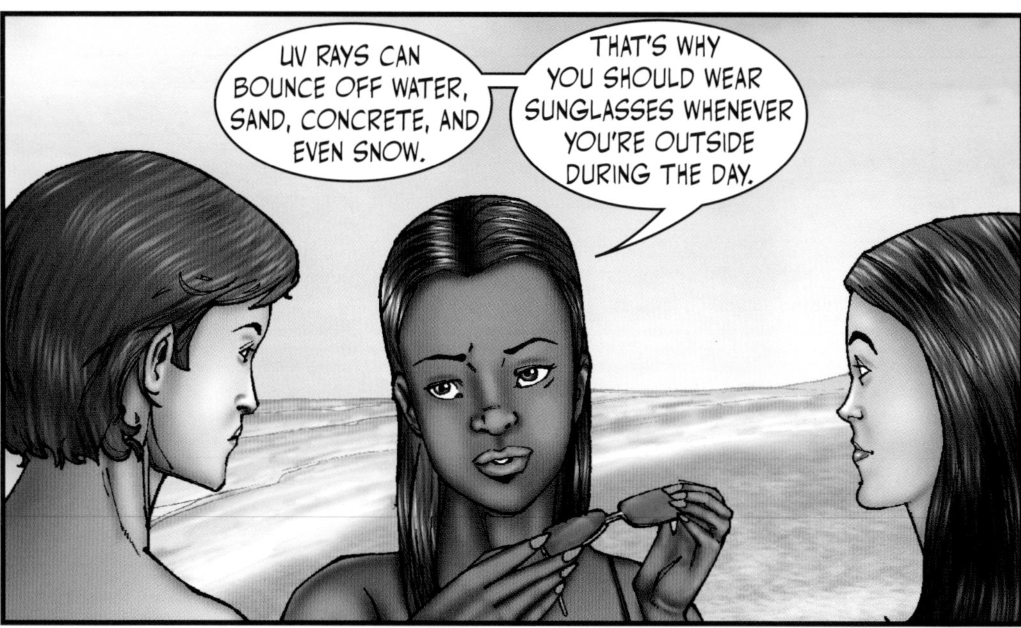
UV RAYS CAN BOUNCE OFF WATER, SAND, CONCRETE, AND EVEN SNOW.
THAT'S WHY YOU SHOULD WEAR SUNGLASSES WHENEVER YOU'RE OUTSIDE DURING THE DAY.

IF UV RAYS ARE SO DANGEROUS, SHOULDN'T WE STAY INSIDE DURING THE DAY?

I WAS WORRIED ABOUT THAT, TOO, WHEN I FIRST LEARNED ABOUT THIS STUFF. BUT THERE IS AN INVISIBLE BARRIER AROUND EARTH CALLED THE OZONE LAYER.

"THE OZONE LAYER ACTS LIKE A SAFETY NET AROUND EARTH.
"IT MAKES SURE THAT ONLY A SMALL AMOUNT OF DANGEROUS UV RAYS REACHES THE SURFACE."

WHOA. GOOD THING WE HAVE THE OZONE LAYER, HUH?

YES, BUT...
IF WE'RE NOT CAREFUL, WE MIGHT NOT ALWAYS HAVE AN OZONE LAYER.

WHAT? WHAT DO YOU MEAN, SARA?

"IN THE 1920S, CHEMICALS CALLED CHLOROFLUOROCARBONS WERE INVENTED.
"CHLOROFLUOROCARBONS, OR CFCS, WERE USED IN AIR CONDITIONERS, REFRIGERATORS, AND AEROSOL SPRAY CANS.

"IN 1979, SCIENTISTS WORKING FOR THE BRITISH ANTARCTIC SURVEY DISCOVERED THAT THE OZONE LAYER OVER ANTARCTICA WAS GETTING THINNER.

"THE ATMOSPHERE OVER THE SOUTH POLE WAS MISSING OZONE GAS.

"IN IMAGES TAKEN FROM SPACE, THIS THINNING OF THE OZONE LAYER OVER ANTARCTICA LOOKED LIKE A HOLE IN EARTH'S SAFETY NET.

"UNFORTUNATELY, BY THE TIME SCIENTISTS DISCOVERED THAT CFCS WERE HAVING THIS EFFECT ON THE OZONE LAYER, THESE CHEMICALS HAD BEEN IN USE FOR DECADES.

"CFCS ARE ALSO FOUND IN POLLUTION FROM AUTOMOBILES, FACTORIES, AND POWER STATIONS.

"WHEN CFCS REACH THE OZONE LAYER, THEY ARE BROKEN APART BY UV RAYS.
"THIS PROCESS RELEASES A CHLORINE ATOM, WHICH CAN BREAK DOWN AND DESTROY OZONE.
"BEFORE LEAVING THE ATMOSPHERE, A SINGLE CHLORINE ATOM CAN DESTROY MORE OZONE THAN CAN BE PUT BACK."

BY THE YEAR 2000, THE HOLE OVER ANTARCTICA WAS THREE TIMES AS LARGE AS THE UNITED STATES.
IF THE OZONE LAYER IS COMPLETELY DESTROYED, THE UV RAYS FROM THE SUN COULD CAUSE WORLDWIDE **DAMAGE.**

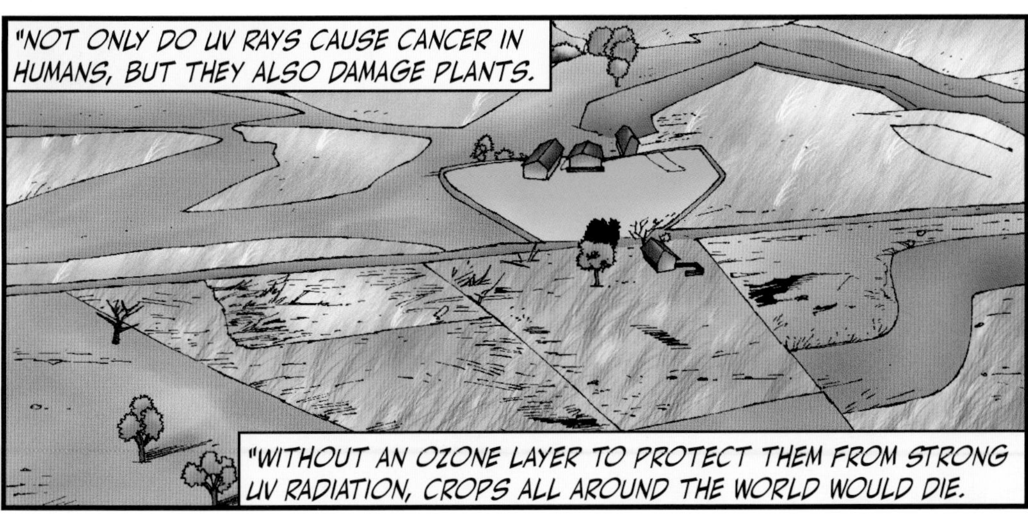
"NOT ONLY DO UV RAYS CAUSE CANCER IN HUMANS, BUT THEY ALSO DAMAGE PLANTS.
"WITHOUT AN OZONE LAYER TO PROTECT THEM FROM STRONG UV RADIATION, CROPS ALL AROUND THE WORLD WOULD DIE.

"THE UV RAYS WOULD ALSO KILL THE PLANTS IN THE OCEANS.
"THE FISH THAT EAT THESE PLANTS WOULD DIE, AND THE ANIMALS THAT EAT THE FISH WOULD SOON DIE, TOO.

"WHEN UV RAYS REACH THE SURFACE OF EARTH, THEY CAN ACTUALLY CREATE OZONE CLOSE TO THE GROUND.

"WE WANT OZONE IN OUR ATMOSPHERE, BUT IF OZONE IS CREATED CLOSE TO THE GROUND, IT CAN BE HARMFUL TO HUMANS.

"WHEN OZONE MIXES WITH AIR POLLUTION, IT CREATES SMOG, WHICH CAN MAKE BREATHING VERY DIFFICULT."

"SMOG AND OTHER KINDS OF AIR POLLUTION TRAP HEAT FROM THE SUN AND KEEP IT CLOSE TO EARTH'S SURFACE.
"EARTH STARTS TO HEAT UP AND TEMPERATURES RISE, JUST LIKE WHAT HAPPENS INSIDE OF A GREENHOUSE. THAT'S WHY IT'S CALLED THE GREENHOUSE EFFECT.
"THIS RISE IN TEMPERATURE CAUSED BY THE GREENHOUSE EFFECT IS CALLED GLOBAL WARMING.

"AS TEMPERATURES INCREASE AROUND THE WORLD, ENTIRE **ENVIRONMENTS** CAN BE AFFECTED.
"RIVERS AND LAKES COULD DRY UP. FARMLAND COULD TURN INTO DESERT.

"IF EARTH GETS TOO HOT, THE POLAR ICE CAPS WILL MELT.

"THIS WOULD CAUSE SEA LEVELS ACROSS THE GLOBE TO RISE.
"COASTAL CITIES, LIKE NEW YORK CITY AND SAN FRANCISCO, WOULD BE FLOODED."

WOW, SARA! I KNEW THINGS WERE BAD, BUT THAT SOUNDS TERRIBLE!
WHAT CAN WE DO TO MAKE SURE THAT DOESN'T HAPPEN?

"COUNTRIES AROUND THE WORLD HAVE PROMISED TO STOP USING CFCS.*
"THEY HAVE ALSO AGREED TO FIND CLEANER ALTERNATIVES FOR OTHER CHEMICALS AND MATTER THAT CAN HARM THE OZONE LAYER OR CONTRIBUTE TO THE GREENHOUSE EFFECT.
*SINCE 1987, 191 COUNTRIES HAVE SIGNED THE MONTREAL PROTOCOL, AN INTERNATIONAL TREATY MEANT TO PROTECT THE OZONE LAYER.

"WE CAN ALL HELP OUT BY CARPOOLING, RIDING OUR BICYCLES, OR TAKING PUBLIC TRANSPORTATION WHENEVER POSSIBLE TO REDUCE POLLUTION FROM CAR **EXHAUST**.

"REUSING AND RECYCLING CAN ALSO HELP GET RID OF POLLUTANTS THAT CAUSE THE GREENHOUSE EFFECT."

IF EVERYONE AROUND THE WORLD STARTS WORKING TOGETHER . . .
. . . SCIENTISTS BELIEVE THAT THE OZONE LAYER COULD BE COMPLETELY REPAIRED IN THE NEXT 50 YEARS.

UNTIL THE OZONE LAYER IS FIXED, THOUGH, WE SHOULD ALWAYS MAKE SURE TO PROTECT OURSELVES FROM THE SUN.

THAT'S RIGHT. AND TOMMY SHOULD REMEMBER TO PAY ATTENTION IN CLASS!
I WILL, IF YOU REMEMBER TO LAY OFF! BELIEVE ME, I'LL BE WEARNG SUNSCREEN FROM NOW ON.
THE END.

Facts About Global Warming

1. Were it not for the greenhouse effect, Earth would be at least 60° F (30° C) cooler and we could not live here.

2. Some scientists say that in about 80 years, the world will be 6.5° F (3.5° C) warmer than it is now.

3. During the last ice age, Earth's average temperatures were only about 9° F (5° C) cooler than they are today.

4. In 1997, during the UN Conference in Kyoto, Japan, leaders from many countries agreed to cut the amount of greenhouse gases that industries, or businesses, make.

5. Since preindustrial times, the level of carbon dioxide in the air has risen 36 percent and the level of methane has risen 148 percent.

6. Ninety percent of Earth's ozone lies within the ozone layer.

7. Scientists think that the global average sea level will rise by 7 to 24 inches (18–61 cm) by 2100.

8. Total U.S. emissions of greenhouse gases have risen by 16.3 percent from 1990 to 2005 and are expected to keep rising.

9. Three-quarters of the greenhouse gases the United States creates come from burning fossil fuels.

10. Out of 10 million air molecules, only 3 are ozone.

GLOSSARY

ALTERNATIVES (ol-TER-nuh-tivz) New or different ways.

ATMOSPHERE (AT-muh-sfeer) The gases around an object in space. On Earth this is air.

ATOMS (A-temz) The smallest parts of elements.

BARRIER (BAR-ee-er) Something that blocks something else from passing.

CHEMICALS (KEH-mih-kulz) Matter that can be mixed with other matter to cause changes.

DAMAGE (DA-mij) Harm.

ENVIRONMENTS (en-VY-ern-ments) All the living things of places.

EXHAUST (ig-ZOST) Smoky air made by setting fire to gas, oil, or coal.

MOLECULES (MAH-lih-kyoolz) Atoms that are joined together.

PARTICLES (PAR-tih-kulz) Small pieces of matter.

SOLAR RADIATION (SOH-lur ray-dee-AY-shun) Rays of light, heat, or energy that spread outward from the Sun.

TEMPERATURES (TEM-pur-cherz) How hot or cold things are.

ULTRAVIOLET RAYS (ul-truh-VY-uh-let RAYZ) Rays given off by the Sun that are dangerous to our skin and eyes.

INDEX

WEB SITES

Due to the changing nature of Internet links, PowerKids Press has developed an online list of Web sites related to the subject of this book. This site is updated regularly. Please use this link to access the list:

www.powerkidslinks.com/ged/warming/